MW01380013

AURORA BOREALIS
A Photo Memory

Published by:

Todd Communications

611 E. 12th Ave., Suite BPM
Anchorage, Alaska 99501-4603
Phone: 907-274-TODD (8633)
Fax: 907-929-5550
e-mail: sales@toddcom.com
WWW.ALASKABOOKSANDCALENDARS.COM

Editor: Flip Todd
Design: Vered R. Mares, **Todd Communications**
Copy writers: Stan Wise & Liz Russo

Printed in China through **Alaska Print Brokers**, Anchorage, Alaska.

Second Edition
First printing July, 2010
Second printing April, 2013
Third printing October, 2018

ISBN: 978-1-57833-458-2
Library of Congress Control Number: 2010901299

Photography:
Naoki Aiba	Azuma Hotta	Norio Matsumoto	Robert Siciliano
Dennis Anderson	Wayne Johnson	Daryl Pederson	Yasuyoshi Tanaka
Larry Anderson	Matsuo Kawashima	Hugh Rose	
Patrick Endres	Didier Lindsey	Todd Salat	
Ulrike Haug	Bob Martinson	John Sandy	

Dust jacket cover: A rarely seen, vivid blue aurora lights up the sky near Homer, Alaska on October 29, 2003. *Photo © Dennis Anderson*

Pages 2-3: A purple aurora shines quietly in a starry sky over Slope Mountain at the start of the Arctic coastal plain. The trans-Alaska pipeline is in the foreground. *Photo © Patrick Endres*

AURORA
BOREALIS

A Photo Memory

An unusual red aurora appears just after sunset over an oil flow line at the Endicott Oil Field. The Endicott Field is the third largest oil field on Alaska's North Slope. *Photo © Larry Anderson*

The bright green aurora illuminates an oil line adorned with ice over Alaska's barren North Slope. *Photo © Larry Anderson*

A brilliant red and green auroral curtain undulates over Alaska's North Slope oil fields. *Photo © Larry Anderson*

The northern lights highlight the sky over the Arctic coastal plain at dusk. Exhaust from Pump Station Three of the trans-Alaska pipeline is illuminated by the station lights, and a passing truck lights up the James Dalton Highway. *Photo © Patrick Endres*

As the sun sets a nearly linear aurora lights the way over the North Slope haul road (now known as the James W. Dalton Highway). The 414-mile-long highway was built in 1974 to support the construction of the trans-Alaska pipeline. *Photo © Didier Lindsey*

Snowden Mountain in Alaska's Brooks Range seems to be on fire with the green light of the aurora borealis.
Photo © Patrick Endres

A dramatic aurora burns like a candle flame over Snowden Mountain in the Brooks Range on the night of March 20, 2003. A real flame would have been more welcome to the photographer on this night when temperatures were about 20 degrees below zero. *Photo © Patrick Endres*

Photographed in Atigun Pass on March 10, 2007, this aurora danced all night long. The light from the aurora reflects off the snowfields and projects a light that is bright enough to read a book by. *Photo © Azuma Hotta*

A red and green aurora lights up the sky over snow-covered James Dalton Mountain near Atigun Pass in the Brooks Range. Atigun pass is the highest and northernmost highway pass in Alaska. *Photo © Todd Salat*

For two minutes this green aurora brightened up the night sky at Atigun Pass in the heart of the Brooks Range. After swirling in a whirlpool motion, the aurora dispersed and the stars were visible again. *Photo © Azuma Hotta*

Right: The northern lights stalk over the Brooks Range like an ethereal brontosaurus at Atigun Pass. At 4,800 feet, Atigun Pass is the highest point along the trans-Alaska pipeline. *Photo © Todd Salat*

James Dalton Mountain in the Brooks Range is illuminated by the rising moon and seems to erupt as a rare red aurora shines brightly above it on a late October night in 2001. *Photo © Todd Salat*

A vibrant green aurora streaks over the Koyukuk River basin in the Brooks Range on March 17, 2005. The light from a first quarter moon illuminates the snow-covered river. *Photo © Patrick Endres*

A green aurora stretches over the Brooks Range like the fingers of a ghostly hand. The Brooks Range stretches east to west across northern Alaska and into Canada. *Photo © Larry Anderson*

A spectacular green aurora brightens the night sky over a tent in the Brooks Range. *Photo © Hugh Rose*

A truck makes its way along Alaska's James Dalton Highway beneath the splendor of the northern lights. *Photo © Hugh Rose*

A curtain of purple, red and yellow light dances above the boreal forest of interior Alaska. The boreal forest is a nearly continuous belt of coniferous trees across North America and Eurasia just below the Arctic Circle. It accounts for about one third of earth's total forest area. *Photo © Hugh Rose*

The northern lights appear to stream from a mountain just south of the James Dalton Highway. Even though the lights seem to touch the mountain, they are actually many miles above the ground in the atmosphere. *Photo © Patrick Endres*

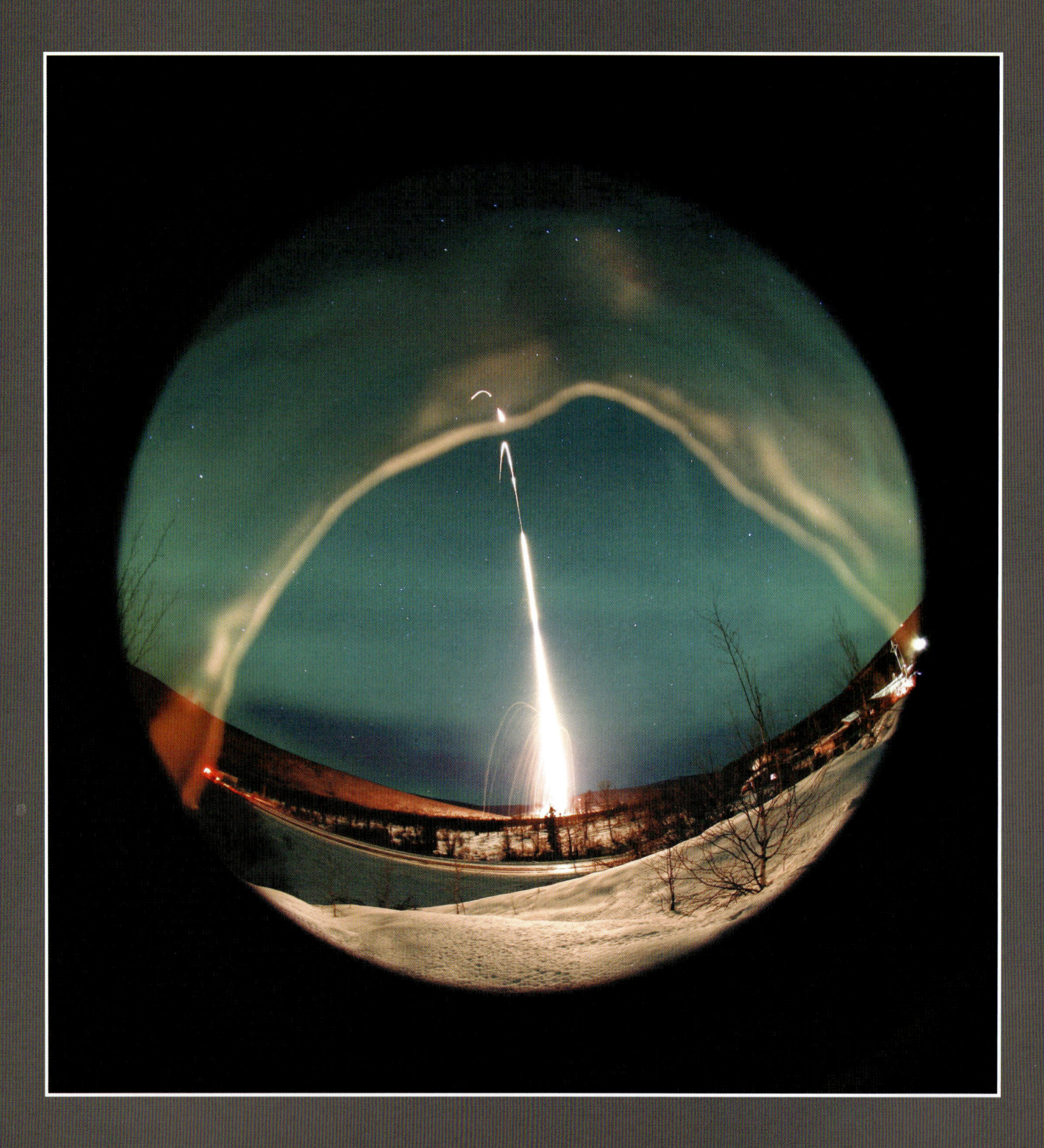

With auroras lighting up the night sky, rockets were launched to a height of 210 km from the University of Alaska Poker Flat Research Range located in Chatanika, Alaska during February 2008.   *Photo © Matsuo Kawashima*

During an active green aurora, this rocket was launched from the Poker Flat Research Range operated by the University of Alaska's Geophysical Institute.   *Photo © Yasuyoshi Tanaka*

A campfire along the Chena River 50 miles north of Fairbanks and a rare, blue aurora make a cold March evening glow with warmth. *Photo © Dennis Anderson*

The aurora borealis rises like a mountain peak over a rustic cabin five miles south of Chena Hot Springs, a resort community built around a natural hot spring.
Photo © Todd Salat

The northern lights shine serenely over historic St. Paul's Church in Eagle, Alaska. The photographer lit the steeple with a handheld flash while the camera's shutter was locked open. *Photo © Todd Salat*

A brilliant green aurora takes the shape of a piano as it twists and swirls over the headwaters of the Fortymile River near Chicken, Alaska. *Photo © Dennis Anderson*

At sun up on March 23, 2001, the purple, green and blue lights of the aurora combined with the orange glow of sunrise to create a beautiful image. This night the aurora activity was continuous at Cleary Summit, about 20 miles north of Fairbanks. *Photo © Naoki Aiba*

From the Big Dipper, this red corona began to break up. At Cleary Summit 20 miles north of Fairbanks, spectacular red auroras could be seen throughout the night on March 23, 2001. *Photo © Naoki Aiba*

At Cleary Summit, 20 miles north of Fairbanks, on March 23, 2002 aurora activity began before darkness fell. Despite the full moon, this partially formed corona was visible and took on the image of a bird stretching its' wings. *Photo © Naoki Aiba*

Just before moonrise on Cleary Summit, these auroras dance on a fresh blanket of snow, creating a mystical world. *Photo © Ulrike Haug*

A startling green aurora loops over Alaska's boreal forest, while the city lights of Fairbanks illuminate the clouds.
Photo © Patrick Endres

A historic gold dredge is silhouetted by a red and green aurora near Fairbanks, Alaska.
Photo © Hugh Rose

A red aurora burns brightly over spruce and birch trees in Fairbanks, Alaska. The yellow is caused by the clouds being illuminated from the city lights of Fairbanks. *Photo © Patrick Endres*

The rays of this corona run straight as train tracks, yet seem to come together off in the distance. This illusion causes the viewer to look twice at the handle of the Big Dipper, as it appears to be bent. *Photo © Ulrike Haug*

In late October 2002 over the hills beyond Fairbanks, this pastel aurora twist and bent in the starry night. Sound carries well in the cold, dry winter air of the Interior. Beneath this aurora a moose cow and her calf stroll by in search of food, their hooves crunching in the snow.
Photo © Ulrike Haug

Despite the brightness of this corona, the Big Dipper is still visible through the aurora. The shape of the aurora is similar to that of a bird, swooping through the sky just south of Delta Junction, Alaska. On this night, the light from the aurora was bright enough to make shadows on the snow.
Photo © John Sandy

Clouds part to reveal a magenta corona streaming through the light of a full moon near Black Rapids Glacier in the Alaska Range. *Photo © Todd Salat*

The northern lights swirl over the Alaska Range seen from Canwell Glacier. The mountains are lit by a full moon on this very cold night in early April 2000. *Photo © Patrick Endres*

A thin purple aurora shimmers over the trans-Alaska pipeline as it enters the ground on its southbound journey at Isabel Pass in the Alaska Range. *Photo © Todd Salat*

The full moon shines through a corona north of Paxson, along the Richardson Highway.
Photo © Todd Salat

Looking east from Denali State Park at dawn on March 24, 2002, the colors of this aurora mix with a sunrise.
Photo © Dennis Anderson

Left: A blue aurora dances above Denali in the Alaska Range during winter. The color of this aurora mirrors the cold, long winter nights that characterize interior Alaska.
Photo © Norio Matsumoto

A massive red and green aurora dances above Mount McKinley. The Chulitna River is frozen in the foreground, and the bright dot is Jupiter setting over Ruth Glacier. *Photo © Todd Salat*

The reds, greens and yellows from this aurora light up a dark winter night near Mt. Foraker in the Alaska Range. *Photo © Norio Matsumoto*

33

Left: A teal aurora smiles down on a candle lit tent near Whistler Glacier in the Alaska Range. The aurora appears to be trying to tease the occupants of the tent into coming out to play. *Photo © Norio Matsumoto*

Below: This teal colored aurora snakes across the sky in the Alaska Range near the Whistler Glacier during winter. A silhouette of Denali is visible in the foreground.
Photo © Norio Matsumoto

Above: With the sound of ice cracking during spring thaw on the Delta River just north of Paxson, Alaska, this green aurora was captured along with the comet Ikeya-Zhang. The arc of light from the aurora silhouettes the Alaska Range.
Photo © Naoki Aiba

Right: The moon shines through clouds beneath a corona in this image taken with a fisheye lens on Deshka Landing Road off Alaska's Parks Highway. A corona occurs when an aurora appears directly over the observer. Though the rays appear to converge toward the center of the aurora, they are actually parallel. *Photo © Wayne Johnson*

An aurora surges across the shores of Christensen Lake near Talkeetna, Alaska. A full moon shines through the red lights on the fringe of the aurora. *Photo © Dennis Anderson*

Right: A magnificent red, green and purple aurora dances in front of Cassiopeia over the photographer's cabin in Hatcher Pass, located in Alaska's Talkeetna Range. *Photo © Todd Salat*

Seen through a 180-degrees full-sky lens, a corona explodes from the Big Dipper over a cabin in Hatcher Pass.
Photo © Todd Salat

A rare pink aurora lights up the pre-dawn sky over the Matanuska Glacier on the morning of October 29, 2003. The pink light occurs when sunlight or bright moonlight falls on the upper atmosphere during an aurora. Through a process called resonance scattering, the sunlight enhances the blue-purple light which is usually too faint to be seen.

Photo © Wayne Johnson

Mountain peaks near Palmer, Alaska, appear to erupt with an eerie green light. Although the aurora seems to emanate from the mountains, it is actually occurring many miles above them high in the earth's atmosphere.
Photo © Daryl Pederson

Off of Farm Loop Road in Palmer, Alaska, an aurora spins over a hay field, illuminating the Talkeetna Mountains. Sometimes the center of an auroral oval will fluctuate farther south, allowing the center to be seen from a lower latitude.
Photo © Bob Martinson

Above: This aurora danced through a mid-March night near Palmer, Alaska in 2003. The common green color that usually dominates auroras is outshone by unusual pinks and purples, some that were strong enough to turn the snow the color of their lights.
Photo © Robert Siciliano

Right: A great sweeping arc of green and purple light over the Chugach Mountains seems to usher in a new millenium on a night in late December 1999.
Photo © Wayne Johnson

Left: The comet Ikeya-Zhang streaks over the Chugach Mountains and through the light of a pink aurora on a calm morning. The Andromeda galaxy appears below the comet.
Photo © Daryl Pederson

The northern lights stream across the Big Dipper over the buildings of Wrangell-St. Elias National Park's historic Kennecott mill town lit with a million-candle-power spotlight. The mill processed copper ore obtained from a nearby mine which operated from 1911 to 1938.
Photo © Todd Salat

On September 27, 2002 a dancing aurora appears just south of Hay River, Northwest Territories, Canada near the Twin Falls Gorge Territorial Park.
Photo © John Sandy

Above: A fisheye lens captures a dazzling red and green display over Eklutna Lake.
Photo © Wayne Johnson

Left: A red aurora highlights the night sky over frost-covered trees at the west end of Eklutna Lake in November 2001.
Photo © Wayne Johnson

Over the Knik River in Southcentral Alaska, a double curtain aurora spins into a circle. The Talkeetna Mountains glow in the distance from the aurora's light. *Photo © Bob Martinson*

Under the light of a full moon, a band of aurora stretches over the Chugach Mountains in the Knik River Valley, near Palmer, Alaska. *Photo © Bob Martinson*

Alpine and celestial majesty are reflected in the smooth waters of the Knik River on November 4, 2001, as a beautiful green aurora illuminates rugged snow-capped peaks. *Photo © Didier Lindsey*

On a calm, autumn evening the northern lights reflect off the waters of the Knik River, stretching from shore to shore. The green auroras outshine the faint orange glow of the Palmer city lights in the distance. *Photo © Robert Siciliano*

Right: A beam of red light appears to shine through the clouds over Eagle River Valley in this magnificent aurora. In fact, the aurora is taking place far above the clouds. The red aurora occurs when the incoming particle stream only has enough energy to penetrate the atmosphere to an altitude of about 120 miles. The electrons in the particle stream collide with oxygen atoms, which emit red light when struck in the higher altitudes. *Photo © Wayne Johnson*

The sun begins to creep over the Chugach Mountains as an amazing purple and pink corona explodes over Flattop Mountain. This display was caused by a coronal mass ejection from the sun the previous day. When the ejection collided with Earth's magnetic field, it sparked some very intense auroras and caused a few radio blackouts. *Photo © Todd Salat*

The northern lights compete with the lights of Anchorage at midnight on August 19, 1999. August is the beginning of the aurora season in Alaska because the sun is too bright at night to see them in May, June and July. *Photo © Todd Salat*

This magnificent corona shines over Anchorage at 7 a.m. on September 30, 2000. The first rays of the sun, while not visible from the ground, fall on the upper atmosphere and, through an effect known as resonance scattering, create the blue and pink colors in the aurora. *Photo © Todd Salat*

Northern lights shine over Anchorage, competing with the bright city lights. It is no competition, however, as the northern lights are brighter than any street lights. As it twists and turns through the sky, the observer has trouble determining where it begins and where it ends. *Photo © Robert Siciliano*

Left: The bright light of a full moon shines through an eerie green aurora over Anchorage in April 1995. *Photo © Wayne Johnson*

Above: A multi-colored aurora appears above the area aptly named Rainbow, Alaska, southeast of Anchorage on the Seward Highway.
Photo © Daryl Pederson

Left: The spruce trees appear to be on fire with a brilliant red aurora over Turnagain Pass.
Photo © Daryl Pederson

Right: The heavens are aflame over Crow Pass during the extreme solar max event of November 2003. The light from this display was bright enough to read by.
Photo © Daryl Pederson

Below: A loop of light wraps itself around the moon above the Portage and Placer River Valleys on a mild April night. *Photo © Daryl Pederson*

Red skies at night shine over Anchorage and Cook Inlet, as seen from Hope, Alaska. *Photo © Daryl Pederson*

Top Left: On March 31, 2001 this colorful aurora shifts north through the sky over Captain Cook State Park on the north-west Kenai Peninsula. *Photo © Dennis Anderson*

In the waning light on the evening of April 30, 2005, this aurora arced low across the horizon across from the beach at Anchor Point, Alaska. *Photo © Dennis Anderson*

Bottom Left: Looking east behind a Russian church in Ninilchik, Alaska, a ray from a purple aurora shoots off into the night sky. On this night the purple rays were not visible to the naked eye but were picked up by the photographic film. *Photo © Dennis Anderson*

Next Page: Taken at midnight on August 12, this aurora hovers over Deep Creek on the Kenai Peninsula. Like this one, auroras taken at twilight often reflect blues and purples from the ambient light that is still in the sky. *Photo © Dennis Anderson*

The skies above the Kenai Peninsula in Southcentral Alaska are awash with the light of a vibrant pink aurora shortly before dawn. *Photo © Daryl Pederson*

This aurora combines with the lingering light of the midnight sun and the flames from a distant forest fire just outside of Homer, Alaska. Despite all this light in the sky, the Big Dipper appears prominently through it all. *Photo © Dennis Anderson*

From a ridge five miles out of Homer, Alaska the colors of this aurora combine with the first rays of sunrise on an August morning. *Photo © Dennis Anderson*

Looking southeast from Homer, this intense red aurora began at sunset and lasted for hours into the evening. *Photo © Dennis Anderson*

Looking like an approaching thunderstorm, this aurora takes over the night sky on September 30, 2002, east of Peace River, Alberta, Canada. *Photo © John Sandy*